Order Up!

When evaluating an expression, there is a certain order to follow.

Example

Choose the operation that is done first.

4 + (5 × 7) ÷ 7 − 3   5 × 7 or 7−3   4 + (5 × 7) ÷ 7 − 3

- Work inside the **parentheses**.
- **Multiply** and **divide** from left to right.
- **Add** and **subtract** from left to right.

So, **5 × 7** is done first.

4 + 35 ÷ 7 − 3
4 + 5 − 3
9 − 3
6

Choose the operation that is done first.

1) 4 × 3 + 6 ÷ 2   4 × 3 or 3 + 6
2) 9 ÷ (3 + 3) × 4   9 ÷ 3 or 3 + 3
3) (2 × 5 − 3) + 7   5 − 3 or 2 × 5
4) 24 ÷ 8 + 6 × 2   6 × 2 or 24 ÷ 8
5) 3 × (8 ÷ 4)   8 ÷ 4 or 3 × 8
6) 6 × 3 − 10 ÷ (3 − 2)   3 − 2 or 6 × 3
7) 3 + (3 + 6) × 2   3 + 3 or 3 + 6
8) 5 − 3 + 6 − 3 − 2   3 − 2 or 5 − 3
9) 2 × 5 + 9 ÷ (9 ÷ 3)   9 ÷ 3 or 2 × 5
10) 5 − 8 ÷ 4 + (3 × 8)   8 ÷ 4 or 3 × 8
11) (6 × 3 × 5) + 4 × 3   4 × 3 or 6 × 3
12) 3 + (6 × 2 ÷ 24 ÷ 8) − 9   6 × 2 or 24 ÷ 8

Answer Box

A	B	C	D	E	F
3 + 3	3 − 2	3 + 6	3 × 8	2 × 5	9 ÷ 3
G	H	I	J	K	L
6 × 2	5 − 3	24 ÷ 8	4 × 3	6 × 3	8 ÷ 4

Objective: Evaluate an expression, using the order of operations.

Everything's in Order!

Evaluate the expression.

1. $2 \times 4 + 8 \div 4$
2. $2.2 \div (1 + 1) \times 0.03$
3. $(2 \times 4 - 2) + 7$
4. $30 \div 3 + 4 \times 3$
5. $14 \div 2 + 5 = 2 \times (9 \div 3)$ True or false?
6. $9 \div 9 + 4 \times (0.3 + 0.7)$
7. $4.5 \times 2 - 3 + 2.1$
8. $2.4 \div (3 + 9) \times 0.3$
9. $14 - (3.2 \div 0.8) \div 2$
10. $6 \times 5 - 10 \div (4 - 2)$
11. $(1.4 + 1.2 - 0.4) \div 5$
12. To evaluate the expression $21 \div 3 \times (8 + 4)$, the first operation you perform is addition. True or false?

Remember the order of operations!

Answer Box

A	B	C	D	E	F
25	22	0.44	13	0.06	5

G	H	I	J	K	L
10	8.1	12	False	0.033	True

2 **Objective:** Evaluate an expression, using the order of operations.

V Is for Variable!

Example

Evaluate the expression.

$x + (x + 3)$ for $x = 2$.

To evaluate an expression with a variable:

- First substitute the value for the variable.
- Then follow the order of operations.

So, $2 + (2 + 3)$ equals **7**.

$x + (x + 3)$
$2 + (2 + 3)$
$2 + 5$
7

Evaluate the expression.

For $x = 3$
1. $x + 10$
2. $7 + x$
3. $x + x$

For $y = 8$
4. $11 + y$
5. $y + 18$
6. $(12 + 4) + y$

For $t = 1$
7. $t + 4$
8. $15 + t$
9. $35 + t$

For $p = 4$
10. $(4 + p) + 4$
11. $7 + p + 3$
12. $p + 0$

 +  +  =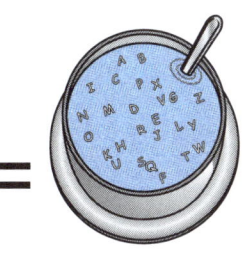

Answer Box

A	B	C	D	E	F
5	12	10	24	14	26

G	H	I	J	K	L
13	19	4	16	6	36

Objective: Evaluate an addition expression with a variable.

Let's Do Decimals!

Evaluate the expression in the rectangle for the value given in the ring.

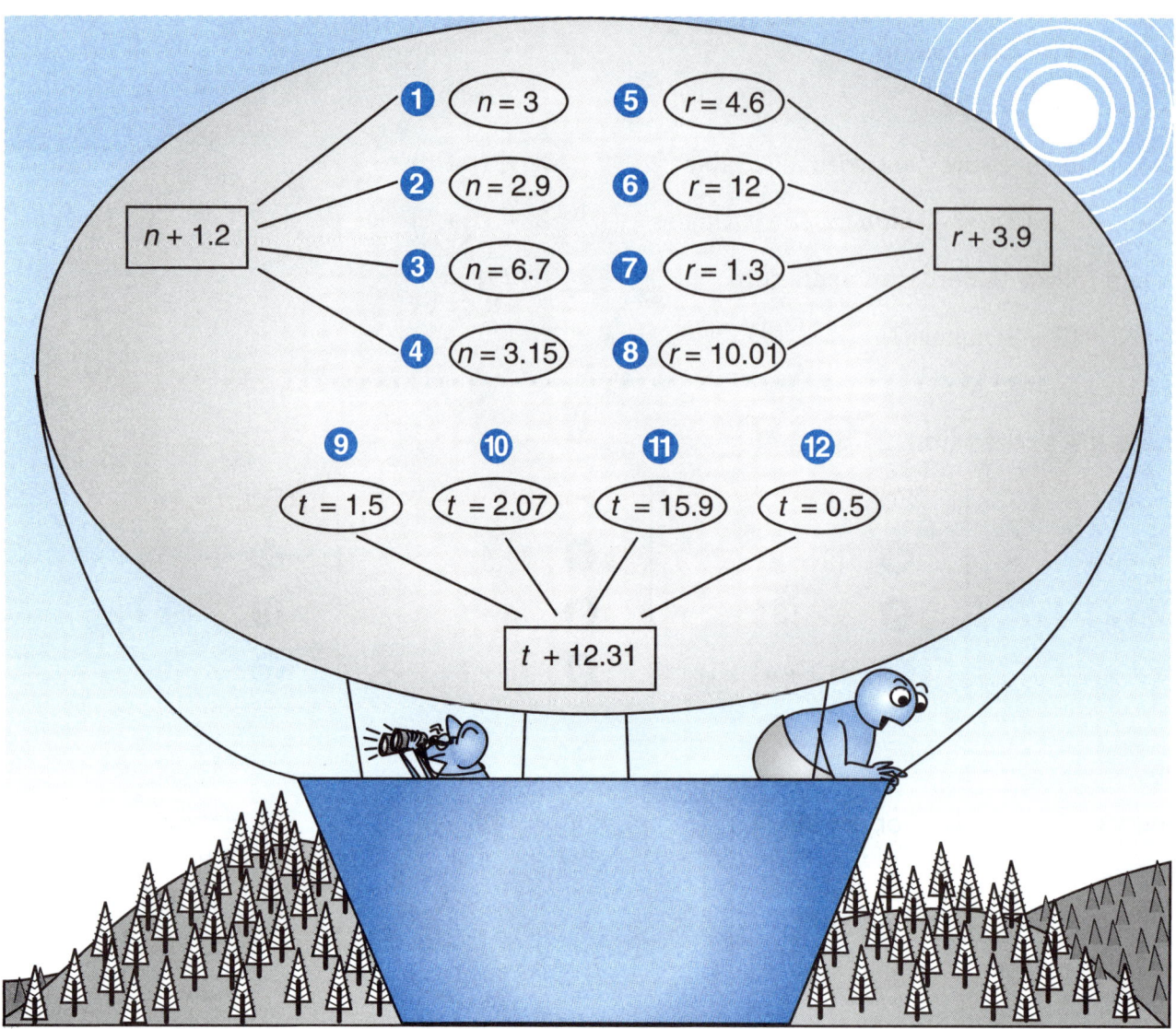

Answer Box

A	B	C	D	E	F
28.21	8.5	13.91	12.81	14.38	5.2
G	H	I	J	K	L
7.9	13.81	15.9	4.1	4.35	4.2

4 Objective: Evaluate an addition expression using decimals.

Take It Away!

> **Example**
>
> **Evaluate the expression.**
>
> $w - (31 - 5)$ for $w = 53$
>
> - First substitute the value of the variable.
> - Then follow the order of operations.
>
> So, $53 - (31 - 5)$ equals **27**.
>
> $w - (31 - 5)$
> $53 - (31 - 5)$
> $53 - 26$
> 27

"You can find the value of a subtraction expression just as you would an addition expression."

Evaluate the expression.

For $b = 13$
1. $14 - b$
2. $b - (3 - 1)$
3. $29 - (b - 12)$

For $d = 28$
4. $120 - d$
5. $d - d$
6. $39 - (36 - d)$

For $f = 110$
7. $f - 100$
8. $f - (f - f)$
9. $(f - 90) - 15$

For $m = 58$
10. $m - 15$
11. $(m - 18 - 10) - 14$
12. $m - (80 - 30)$

Answer Box

A	B	C	D	E	F
11	8	16	0	43	92

G	H	I	J	K	L
110	31	5	28	1	10

Objective: Evaluate a subtraction expression with a variable.

Problem Solving: Using a Formula

Use these formulas to solve the problem.

CONSIDER THIS

Cost of muffins = $m \times \$0.35$, where m = the number of muffins

Number of muffins baked = $t \times 12$, where t = the number of tins of muffins baked

Number of pounds of cookies sold = $c \div 24$, where c = the number of cookies

Cost of cookies = $p \times \$1.50$, where p = the number of pounds of cookies

1 Kara buys 6 dozen cookies. How many pounds does she buy?

2 What is the cost of Kara's 6 dozen cookies?

3 The baker bakes 12 tins of muffins for sale in the morning. How many muffins does he bake?

4 How much will the bakery receive from the sale of muffins if 6 tins are sold?

5 How much will it receive if all 12 tins of muffins are sold?

6 How many muffins would need to be sold to receive $53.90?

7. Seven muffins cost more than $2.10. Yes or no?

8. Three pounds of cookies cost less than $5. True or false?

9. A school orders 24 banana-walnut muffins and 3 lb of oatmeal cookies. How much does the school pay for the order?

10. If the children spent $3.50 at the bakery, could they have bought 3 lb of cookies? Yes or no?

11. Which amount could represent the cost of some muffins, $1.75 or $12.90?

12. Two dozen muffins cost more than $10. True or false?

Answer Box

A	B	C	D	E	F
144	$12.90	$25.20	3	154	$4.50
G	H	I	J	K	L
$50.40	$1.75	Yes	No	False	True

Objective: Solve a problem by substituting values into a simple one-step formula.

7

More Decimal Action!

Evaluate the expression.

1. $y - (3.5 - 1.02)$ for $y = 12.4$
2. $4.02 - 0.5 - d$ for $d = 0.1$
3. $m - (5.3 - 3.01)$ for $m = 15.05$
4. $4.02 - (0.5 - d)$ for $d = 0.1$
5. $m - 5.3 - 3.01$ for $m = 15.05$
6. $y - 3.5 - 1.02$ for $y = 12.4$
7. $m - 3.10 - 5.3$ for $m = 15.05$
8. $p - (13.03 - 0.66)$ for $p = 29.1$
9. $(p - 13.03) - 6.6$ for $p = 29.1$
10. $4.6 - b - b$ for $b = 2.04$
11. $4.6 - (b - b)$ for $b = 2.04$
12. $p - 13.03$ for $p = 29.1$

Answer Box

A	B	C	D	E	F
6.65	9.92	4.6	6.74	7.88	16.73

G	H	I	J	K	L
3.62	0.52	12.76	9.47	16.07	3.42

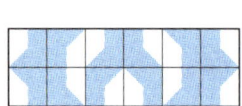

Evaluate the expression.

1. $y - (1.5 - 1.02)$ for $y = 8.4$
2. $4.02 - 1.5 - d$ for $d = 2.1$
3. $(m - 4.3) - 3.01$ for $m = 12.02$
4. $4.02 - (0.5 - d)$ for $d = 0.5$
5. $m - 2.3 - 3.01$ for $m = 13.02$
6. $y - 1.5 - 1.02$ for $y = 9.4$
7. $y - (3.5 - 2.2)$ for $y = 10.4$
8. $6.6 - y - 0.02$ for $y = 6.3$
9. $3.7 - (2.1 - y)$ for $y = 1.5$
10. $8.6 - b - b$ for $b = 4.02$
11. $4.6 - (b - b)$ for $b = 3.07$
12. $p - 11.03$ for $p = 19.1$

Answer Box

A	B	C	D	E	F
6.88	0.42	7.92	4.02	4.71	3.1

G	H	I	J	K	L
8.07	0.28	0.56	9.1	7.71	4.6

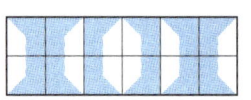

Objective: Evaluate a subtraction expression using decimals.

Model It!

Example

Use a model to help you solve the equation.

$d + 4 = 7$

$d +$ ◎◎◎◎ $=$ ◎◎◎◎◎◎◎

$d + \quad 4 \quad = \quad \quad 7$

To solve the equation, **subtract** the same amount from both sides.

$d + \quad 4 - 4 \quad = \quad \quad 7 - 4$

Subtracting 4 from both sides leaves the variable d alone on one side.

$d =$ ◎◎◎

So, **$d = 3$**.

Use a model to help you solve the equation.

1. $5 + d = 12$
2. $12 + d = 15$
3. $d + 13 = 22$
4. $d + 3 = 9$

5. $46 + d = 54$
6. $d + 33 = 50$
7. $14 + d = 45$
8. $19 + d = 31$

9. $d + 20 = 68$
10. $d + 27 = 38$
11. $43 + d = 84$
12. $100 + d = 135$

Answer Box

A	B	C	D	E	F
$d = 35$	$d = 3$	$d = 31$	$d = 7$	$d = 8$	$d = 48$

G	H	I	J	K	L
$d = 17$	$d = 12$	$d = 6$	$d = 11$	$d = 9$	$d = 41$

Objective: Use a model to represent and solve an addition equation with whole numbers.

Keep Your Balance!

Example

Solve the equation.

$16 + n = 33$

$16 - 16 + n = 33 - 16$

So, **n = 17**.

Remember! To solve an addition equation, you must balance it. What you do to one side, you also do to the other!

Solve the equation.

1. $56 + n = 102$
2. $30 + n = 58$
3. $n + 0 = 36$
4. $n + 10 = 50$
5. $n + 12 = 71$
6. $n + 31 = 43$
7. $n + 33 = 50$
8. $34 + n = 40$
9. If 38 added to a number n equals 45, then is n more than or less than 10?
10. The sum of a number n and 6 equals 30. What is the number?
11. The sum of two addends is 50. If one of the addends is equal to 25, is the second addend more than or less than 20?
12. The sum of a number n plus 27 equals 36. What is the number?

Answer Box

A	B	C	D	E	F
$n = 17$	$n = 24$	$n = 12$	$n = 28$	$n = 59$	More than
G	**H**	**I**	**J**	**K**	**L**
$n = 40$	$n = 46$	$n = 9$	$n = 6$	$n = 36$	Less than

Objective: Solve a 1-step addition equation with whole numbers.

Sensible Subtraction!

Example

Use a model to help you solve a subtraction equation.

$m - 5 = 12$

To solve the equation, **add** the same amount to both sides.

$m - 5 + 5 = 12 + 5$

Adding 5 to both sides leaves the variable m alone on one side.

So, **m = 17**.

Use a model to help you solve the equation.

1. $m - 4 = 31$
2. $m - 7 = 55$
3. $m - 6 = 42$
4. $m - 23 = 19$
5. $m - 32 = 4$
6. $m - 16 = 5$
7. $m - 21 = 25$
8. $m - 7 = 74$
9. $m - 5 = 6$
10. $m - 7 = 3$
11. $m - 61 = 4$
12. $m - 22 = 15$

Answer Box

A	B	C	D	E	F
$m = 65$	$m = 36$	$m = 81$	$m = 21$	$m = 10$	$m = 46$

G	H	I	J	K	L
$m = 11$	$m = 48$	$m = 62$	$m = 37$	$m = 35$	$m = 42$

Objective: Use a model to represent and solve a subtraction equation with whole numbers.

Subtract It!

Example

Solve the equation.

$x - 12 = 15$

$x - 12 + 12 = 15 + 12$

So, **$x = 27$**.

To solve a subtraction equation, you must balance it. What you do to one side, you also do to the other!

Solve the equation.

1. $x - 4 = 10$
2. $x - 12 = 12$
3. $x - 12 = 1$
4. $x - 31 = 30$
5. $x - 8 = 12$
6. $x - 25 = 35$
7. $x - 24 = 12$
8. $x - 7 = 10$

9. If 6 is subtracted from a number x and the result is 10, is x greater than or less than 17?

10. The difference when 13 is subtracted from a number x is 5. What is the number x?

11. If 16 is subtracted from a number x and the difference is greater than 10, is x greater than or less than 26?

12. What number x minus forty-two equals five?

Answer Box

A	B	C	D	E	F
$x = 17$	$x = 60$	$x = 13$	Less than	$x = 18$	$x = 61$
G	H	I	J	K	L
$x = 24$	$x = 47$	Greater than	$x = 20$	$x = 14$	$x = 36$

Objective: Solve a 1-step subtraction equation with whole numbers.

Decimal Dare!

Example

Solve the equation.

4.5 + y = 6.9

    4.5 − 4.5 + y = 6.9 − 4.5

So, **y = 2.4**.

Solve the equation.

1. y + 12.8 = 19.5
2. 3.9 + y = 24.9
3. 1.72 + y = 2.04
4. y + 1.5 = 4.85
5. y + 13.45 = 15.08
6. y + 4.3 = 20.13
7. y + 6.2 = 28.1
8. y + 32.75 = 45.1
9. y + 8.6 = 10.4
10. 2.48 + y = 5.3
11. 8.36 + y = 13.78
12. y + 26.3 = 34.09

You can solve an addition equation with decimals the same way you solve one with whole numbers.

Answer Box

A	B	C	D	E	F
y = 21.0	y = 15.83	y = 6.7	y = 3.35	y = 1.8	y = 0.32
G	H	I	J	K	L
y = 12.35	y = 7.79	y = 2.82	y = 21.9	y = 5.42	y = 1.63

Objective: Solve a 1-step addition equation with decimals.

Problem Solving: Working Backward

Solve the problem.

1. I am thinking of a number. When 34 is added to the number, the sum is 40. What is the number?

2. When I turn over a number card, the number I see is an odd number. If 12 is subtracted from the number, the result is 11. What is the number?

3. If 12 is added to Sparky's age, the result is Spot's age. If 5 is subtracted from Spot's age, the result is Domino's age, which is 9. How old is Sparky?

4. My wand is 12 in. long. If I compare the length of the wand with the height of my hat, the difference is 6 in. My hat is taller than my wand. How many inches tall is my hat?

5. I can perform 3 tricks in 20 min. A fourth trick takes 10 min to perform. At the end of the fourth trick, it is 3:25 P.M. What time did I begin the first trick?

6. The show begins 12 min after the clown act ends. The clown act takes 35 min. If the clown act is scheduled to begin at 1:45 P.M., when will the show begin?

7 If a number is 8 more than another number, the sum of the numbers cannot be less than 9. True or false?

8 My first assistant has 3 more costumes than I have. My second assistant has 7 costumes, 2 more than my first assistant. How many costumes do we have altogether?

9 In the clown act there are some black dogs, white dogs, and cats. There are more white dogs than cats. If there are 5 black dogs and 11 animals altogether, are there more or fewer than 3 cats?

10 I use 12 props for my act. My assistants also use props. Altogether we use 28 props. Do my assistants use more or fewer props than I do?

11 If the difference between two counting numbers is 15, and one addend is less than 15, then the other addend must be greater than 15. True or false?

12 My oldest rabbit is 15 years old. Another rabbit is more than 8 years younger, but not as young as my third rabbit, which is 2 years old. Which could be the combined ages of my 3 rabbits? 18 or 22 years old?

Answer Box

A	B	C	D	E	F
2	True	18	6	22	23
G	H	I	J	K	L
2:55 P.M.	Fewer	More	False	2:32 P.M.	14

Objective: Solve a problem by working backward.

E Is for Equation!

Example

Solve the equation.

$n - 4.51 = 3.7$

$n - 4.51 + 4.51 = 3.7 + 4.51$

So, **n = 8.21**.

Solve the equation.

1. $n - 5.32 = 4.8$
2. $n - 12.8 = 1.02$
3. $n - 4.9 = 5.8$
4. $n - 10.45 = 3.4$
5. $n - 6.31 = 7.2$
6. $n - 11.7 = 9.5$
7. $n - 5.89 = 3.7$
8. $n - 24.06 = 2.83$

9. Sparky weighs 4.5 lb less than another dog, Newton (n). If Sparky weighs 25.75 lb, how many pounds does Newton weigh?

10. What is the value of n in the equation, $n - 17.19 = 8.53$?

11. Which solution cannot be the correct one for the equation, $n - 4.8 = 5.9$? $n = 1.7$ or $n = 10.7$

12. What is the value of n in the equation, $n - 17.6 = 3.8$?

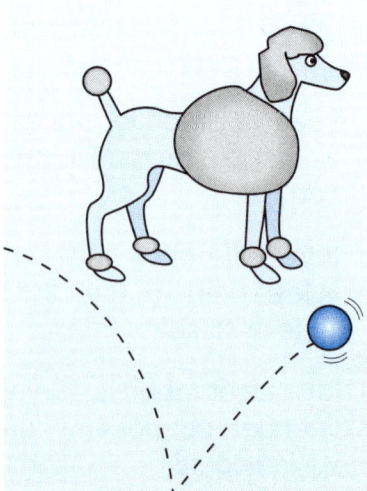

Answer Box

A	B	C	D	E	F
n = 9.59	n = 10.12	n = 13.82	n = 21.2	n = 13.51	n = 30.25
G	H	I	J	K	L
n = 13.85	n = 25.72	n = 21.4	n = 26.89	n = 10.7	n = 1.7

16 Objective: Solve a 1-step subtraction equation with decimals.

Multiplication Models

Example

Use a model to help you solve the equation.

$2 \times n = 8$

Think: How can 8 be divided equally into 2 parts?

$8 \div 2 = 4$ and $2 \times 4 = 8$.

So, **n = 4**.

Use a model to help you solve the equation.

1. $5 \times n = 10$
2. $3 \times n = 33$
3. $9 \times n = 9$
4. $12 \times n = 48$
5. $7 \times n = 42$
6. $10 \times n = 50$
7. $n \times 6 = 48$
8. $n \times 6 = 72$
9. $n \times 7 = 21$
10. $6 \times n = 54$
11. $4 \times n = 28$
12. $n \times 8 = 80$

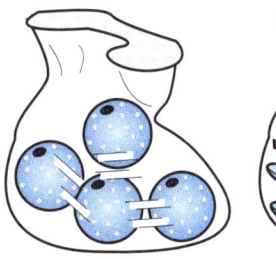

Answer Box

A	B	C	D	E	F
n = 1	n = 7	n = 5	n = 12	n = 4	n = 9
G	H	I	J	K	L
n = 6	n = 3	n = 10	n = 11	n = 8	n = 2

Objective: Use a model to represent and solve a multiplication equation with whole numbers.

Make Mine Multiplication!

Example

Solve the equation.

$s \times \$5 = \$45$

To solve a multiplication equation, divide both sides by the same term.

$$\frac{s \times \$5}{\$5} = \frac{\$45}{\$5}$$

So, **s = 9**.

Solve the equation.

1. $s \times \$12 = \$108$
2. $s \times \$3 = \$48$
3. $s \times \$9 = \$36$
4. $s \times \$24 = \$72$
5. $s \times \$7 = \$70$
6. $s \times \$8 = \$48$
7. $s \times \$5 = \$65$
8. $s \times \$13 = \$91$
9. $s \times \$45 = \$90$
10. $s \times \$12 = \$144$
11. $s \times \$6 = \$90$
12. $s \times \$11 = \$121$

Answer Box

A	B	C	D	E	F
s = 7	s = 11	s = 15	s = 10	s = 12	s = 13
G	**H**	**I**	**J**	**K**	**L**
s = 16	s = 6	s = 2	s = 4	s = 9	s = 3

Objective: Solve a 1-step multiplication equation.

Function Junction!

Use the equation to find the missing value.

1) $b = 3 + a$

a	b
1	4
10	■
12	15

2) $b = 5 \times a$

a	b
1	5
2	10
3	■

3) $b = 11 \times a$

a	b
10	110
3	33
■	121

4) $b = a - 12$

a	b
15	3
25	13
40	■

5) $b = a + 45$

a	b
63	108
■	125
95	140

6) $b = a - 17$

a	b
■	45
21	4
46	29

7) $b = 7 \times a$

a	b
5	35
8	■
9	63

8) $b = 30 + a$

a	b
64	94
■	85
5	35

9) $b = 12 \times a$

a	b
■	144
11	132
9	108

10) $b = a - 68$

a	b
130	62
118	50
■	29

11) $b = a + 11$

a	b
■	21
21	32
33	44

12) $b = 15 \times a$

a	b
5	75
10	■
15	225

Answer Box

A	B	C	D	E	F
$a = 12$	$a = 11$	$b = 28$	$b = 13$	$b = 15$	$a = 62$
G	**H**	**I**	**J**	**K**	**L**
$a = 80$	$a = 10$	$b = 56$	$a = 97$	$b = 150$	$a = 55$

Objective: Using a function machine, name the input or output.

More Machines

Use the equation to find the missing value.

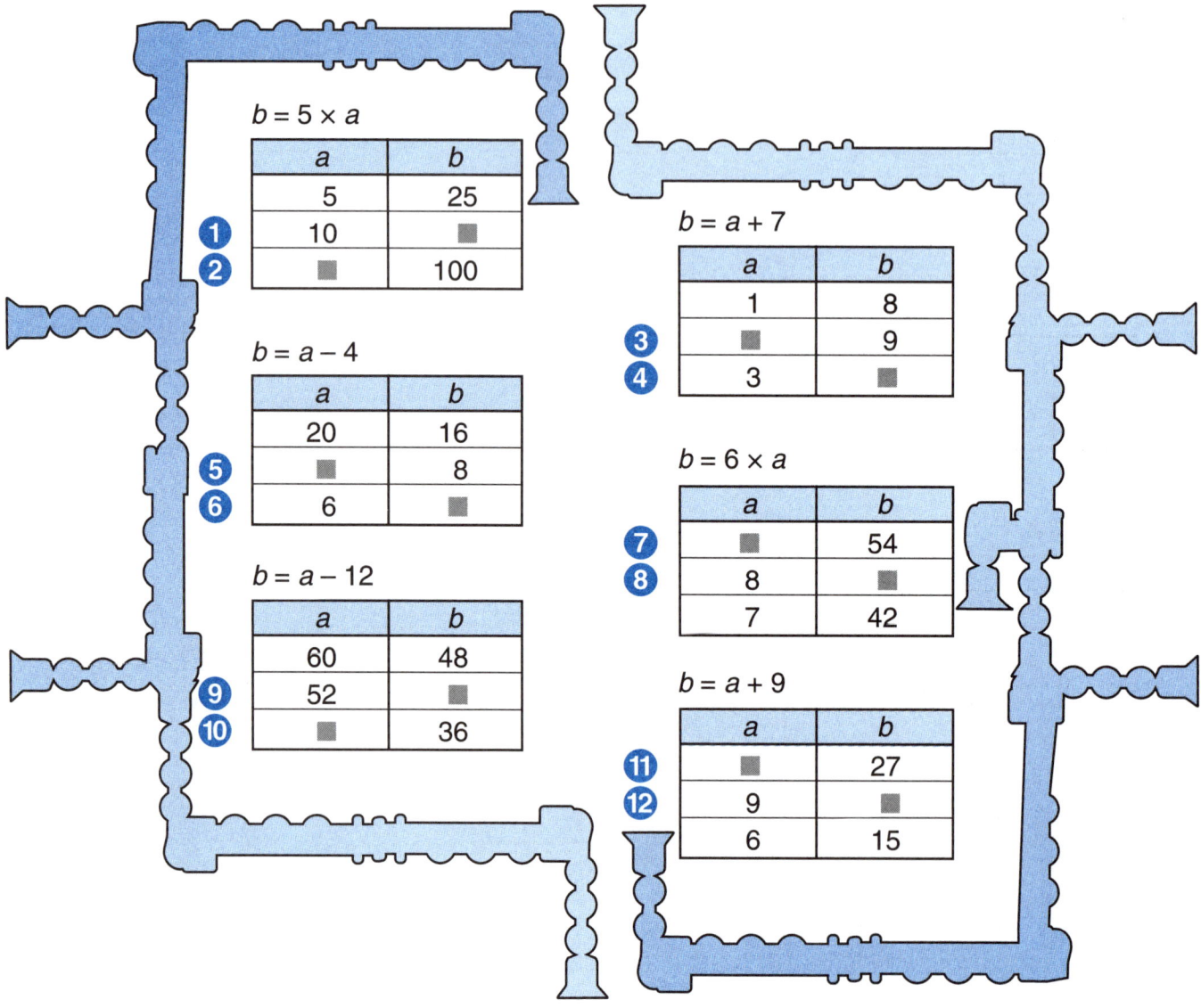

$b = 5 \times a$

a	b
5	25
10	■
■	100

1.
2.

$b = a - 4$

a	b
20	16
■	8
6	■

5.
6.

$b = a - 12$

a	b
60	48
52	■
■	36

9.
10.

$b = a + 7$

a	b
1	8
■	9
3	■

3.
4.

$b = 6 \times a$

a	b
■	54
8	■
7	42

7.
8.

$b = a + 9$

a	b
■	27
9	■
6	15

11.
12.

Answer Box

A	B	C	D	E	F
$b = 50$	$b = 10$	$a = 2$	$a = 18$	$a = 20$	$b = 2$
G	H	I	J	K	L
$a = 48$	$a = 9$	$a = 12$	$b = 40$	$b = 48$	$b = 18$

Objective: Using a function machine, name the input or output.

Fun with Division Equations

Example

Solve the equation.

$x \div 7 = 25$

You can use multiplication to solve a division equation.

$x \div 7 \times 7 = 25 \times 7$

So, **$x = 175$**.

Remember! Multiplication and division are inverse operations.

Solve the equation.

1. $x \div 35 = 2$
2. $x \div 3 = 12$
3. $x \div 8 = 9$
4. $x \div 10 = 6$
5. $x \div 5 = 6$
6. $x \div 9 = 2$
7. $x \div 5 = 3$
8. $x \div 11 = 12$
9. $x \div 7 = 6$

10. The quotient is 24 and the divisor is 4. What is the value of the dividend, x?

11. Four model airplanes are displayed on each of 3 shelves. What is the number of models (x) that are displayed?

12. The quotient is 2 and the divisor is 8. What is the value of the dividend, x?

Answer Box

A	B	C	D	E	F
$x = 36$	$x = 16$	$x = 70$	$x = 15$	$x = 42$	$x = 30$

G	H	I	J	K	L
$x = 132$	$x = 18$	$x = 96$	$x = 60$	$x = 12$	$x = 72$

Objective: Solve a 1-step division equation.

Let's Multiply and Divide!

Choose the value that solves the equation.

1. $r \div 4 = 8$ 32, 20, or 2
2. $r \times 3 = 24$ 72, 36, or 8
3. $r \times 8 = 72$ 4, 8, or 9
4. $r \div 5 = 7$ 5, 35, or 45
5. $r \div 10 = 2$ 5, 20, or 100
6. $r \times 12 = 60$ 5, 9, or 100
7. $r \div 9 = 3$ 9, 18, or 27
8. $6 \times r = 36$ 6, 8, or 9
9. $27 \div r = 9$ 3, 6, or 9
10. $r \times 7 = 28$ 4, 5, or 196
11. $r \div 9 = 4$ 4, 12, or 36
12. $r \times 8 = 56$ 7, 8, or 12

Answer Box

A	B	C	D	E	F
r = 35	r = 4	r = 5	r = 8	r = 20	r = 3
G	H	I	J	K	L
r = 27	r = 32	r = 7	r = 6	r = 9	r = 36

Find the value that solves the equation.

1. $m \div 4 = 5$
2. $m \times 8 = 40$
3. $m \times 12 = 144$
4. $m \div 3 = 3$
5. $m \div 2 = 20$
6. $m \times 11 = 77$
7. $m \div 6 = 9$
8. $m \div 4 = 6$
9. $m \times 8 = 48$
10. $m \times 11 = 22$
11. $m \div 6 = 3$
12. $m \times 4 = 32$

Answer Box

A	B	C	D	E	F
m = 54	m = 2	m = 8	m = 24	m = 40	m = 18
G	H	I	J	K	L
m = 9	m = 20	m = 7	m = 5	m = 12	m = 6

Objective: Solve a 1-step multiplication or division equation.

Get the Point?

Name the coordinates of the point.

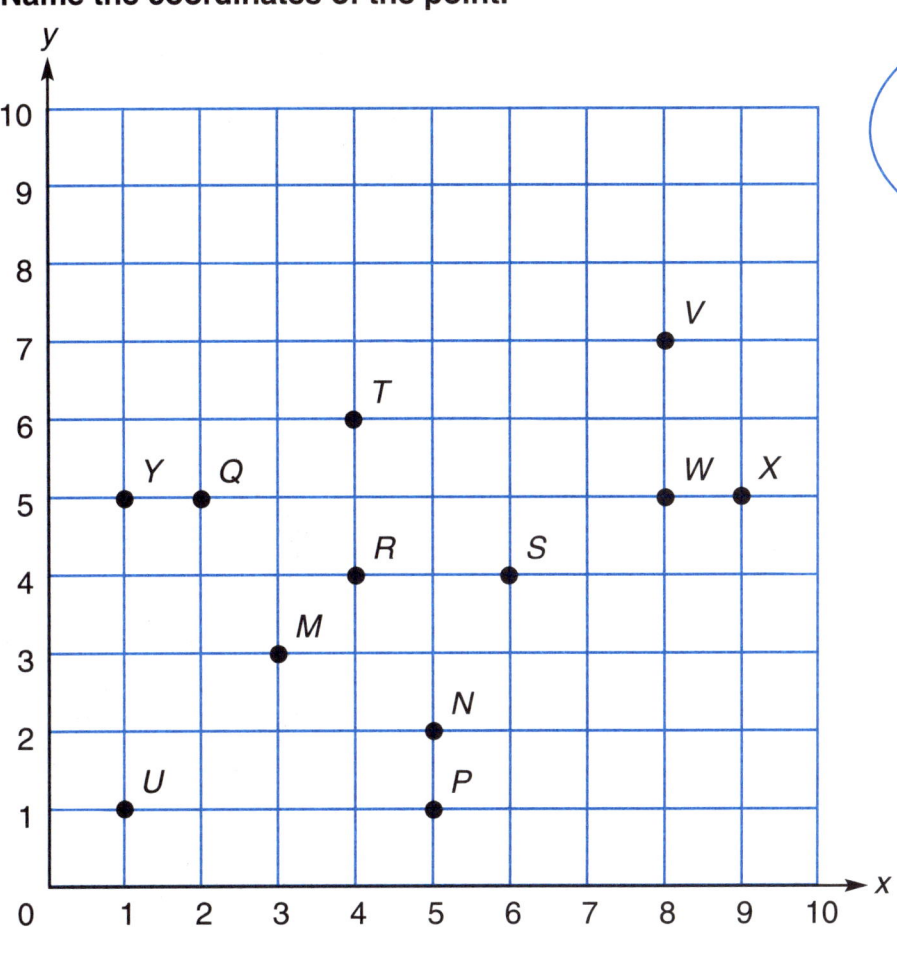

Each ordered pair (x, y) names a value along the x-axis and a value along the y-axis.

1. M
2. N
3. Y
4. P
5. Q
6. R
7. S
8. T
9. U
10. V
11. W
12. X

Answer Box

A	B	C	D	E	F
(2, 5)	(8, 5)	(5, 2)	(4, 4)	(8, 7)	(6, 4)
G	H	I	J	K	L
(1, 1)	(1, 5)	(4, 6)	(9, 5)	(5, 1)	(3, 3)

Objective: Given a point in the first quadrant, name the coordinates.

Moving Around!

Name the coordinates of the point.

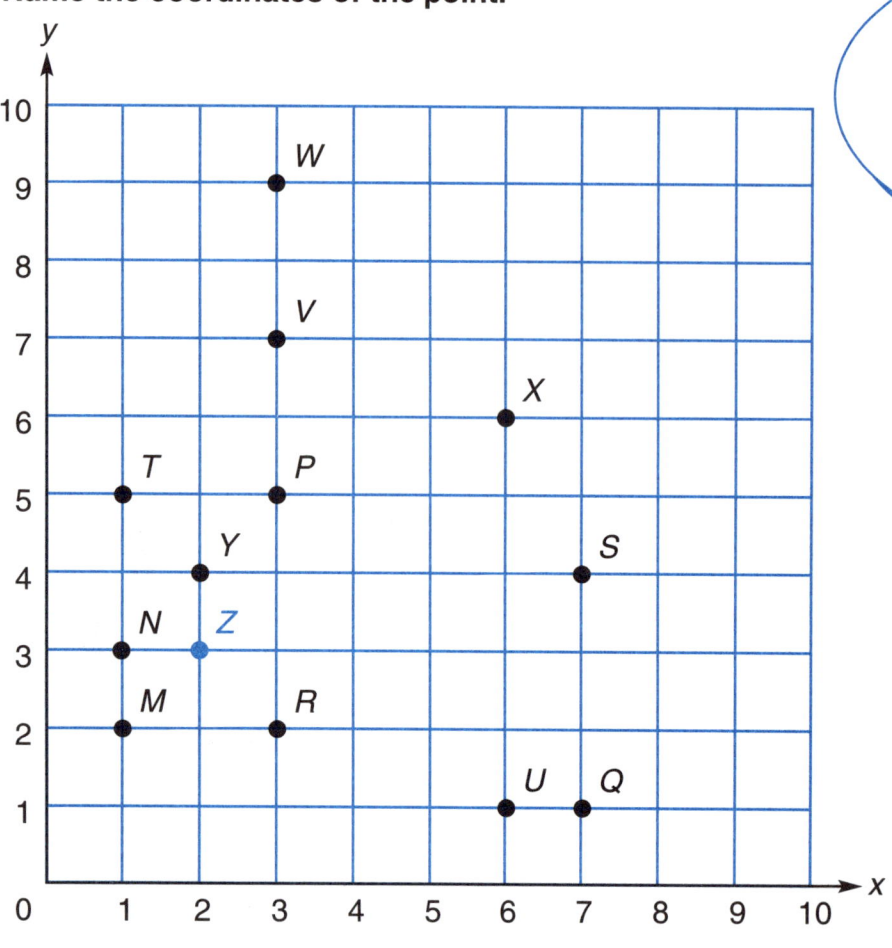

The **ordered pair** (2, 3) names a **point**. To find the point, move 2 units to the right of (0, 0) and then 3 units up. The point is Z.

① M ② N ③ Y ④ P
⑤ Q ⑥ R ⑦ S ⑧ T
⑨ U ⑩ V ⑪ W ⑫ X

Answer Box

A	B	C	D	E	F
(1, 5)	(6, 6)	(3, 9)	(2, 4)	(3, 7)	(7, 4)
G	H	I	J	K	L
(3, 2)	(1, 3)	(6, 1)	(7, 1)	(3, 5)	(1, 2)

24 Objective: Given a point in the first quadrant, name the coordinates.

Let's Graph It!

Name the point on the graph for the ordered pair.

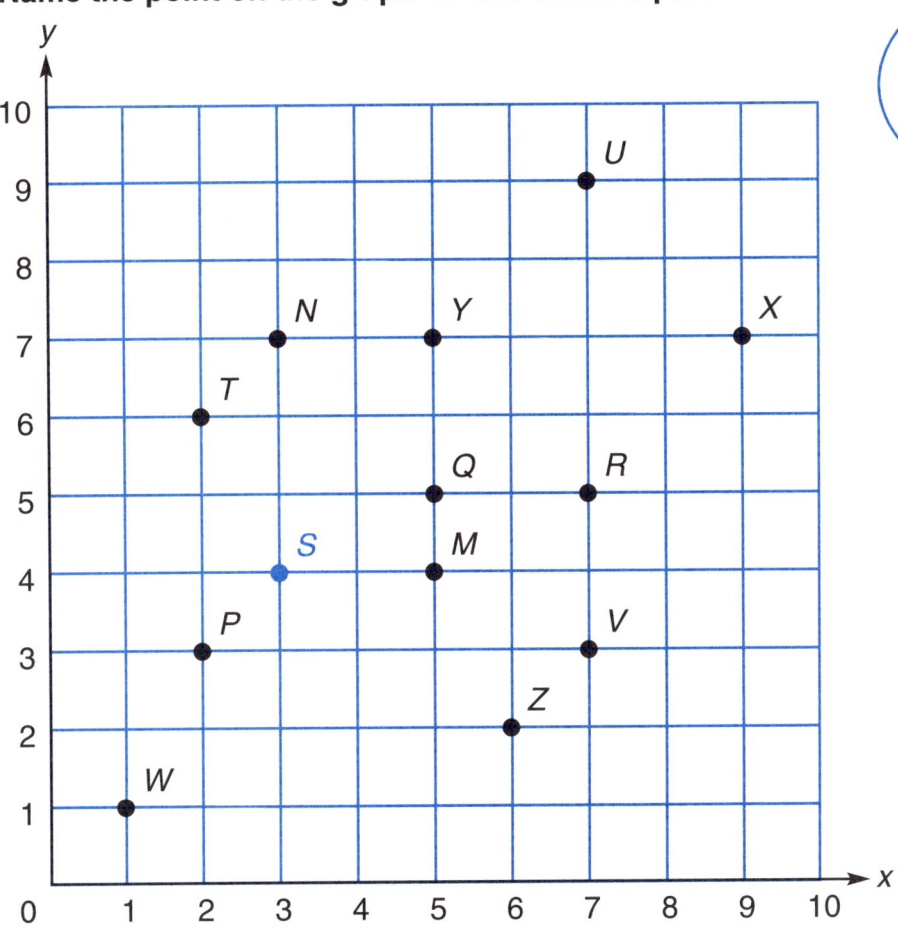

(3, 4) locates point S which is 3 units to the right of (0, 0) and then 4 units up.

① (3, 7)	② (6, 2)	③ (1, 1)	④ (7, 9)
⑤ (2, 3)	⑥ (7, 5)	⑦ (5, 4)	⑧ (2, 6)
⑨ (9, 7)	⑩ (7, 3)	⑪ (5, 5)	⑫ (5, 7)

Answer Box

A	B	C	D	E	F
M	N	P	Q	R	T
G	H	I	J	K	L
U	V	W	X	Y	Z

Objective: Identify the point graphed in the first quadrant for a given ordered pair.

What's Missing?

Find the missing number.

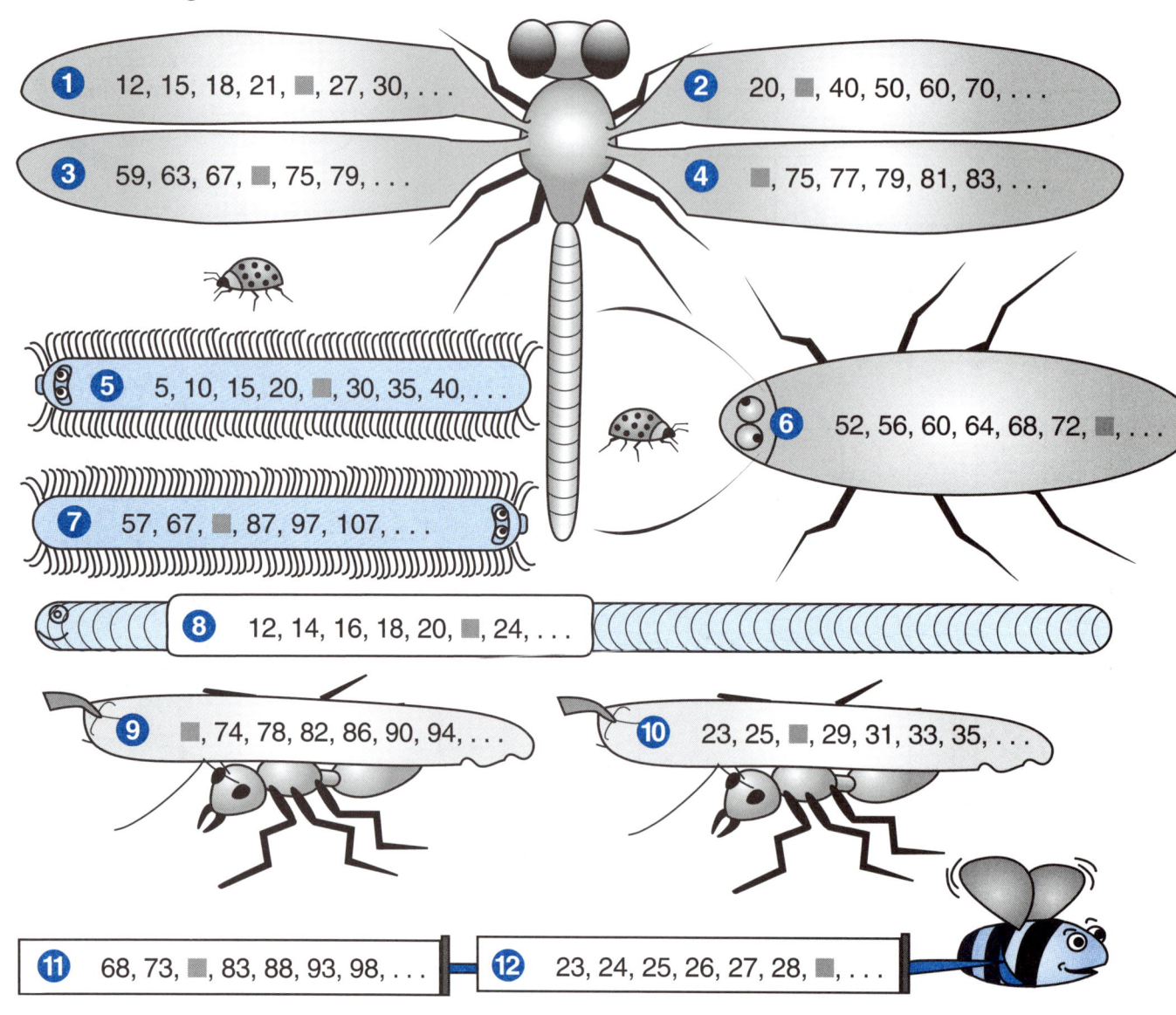

1) 12, 15, 18, 21, ■, 27, 30, . . .
2) 20, ■, 40, 50, 60, 70, . . .
3) 59, 63, 67, ■, 75, 79, . . .
4) ■, 75, 77, 79, 81, 83, . . .
5) 5, 10, 15, 20, ■, 30, 35, 40, . . .
6) 52, 56, 60, 64, 68, 72, ■, . . .
7) 57, 67, ■, 87, 97, 107, . . .
8) 12, 14, 16, 18, 20, ■, 24, . . .
9) ■, 74, 78, 82, 86, 90, 94, . . .
10) 23, 25, ■, 29, 31, 33, 35, . . .
11) 68, 73, ■, 83, 88, 93, 98, . . .
12) 23, 24, 25, 26, 27, 28, ■, . . .

Answer Box

A	B	C	D	E	F
77	27	76	22	25	78

G	H	I	J	K	L
24	73	29	30	70	71

26 Objective: Identify a missing number in an addition number series.

Subtraction Sequences

Find the missing number.

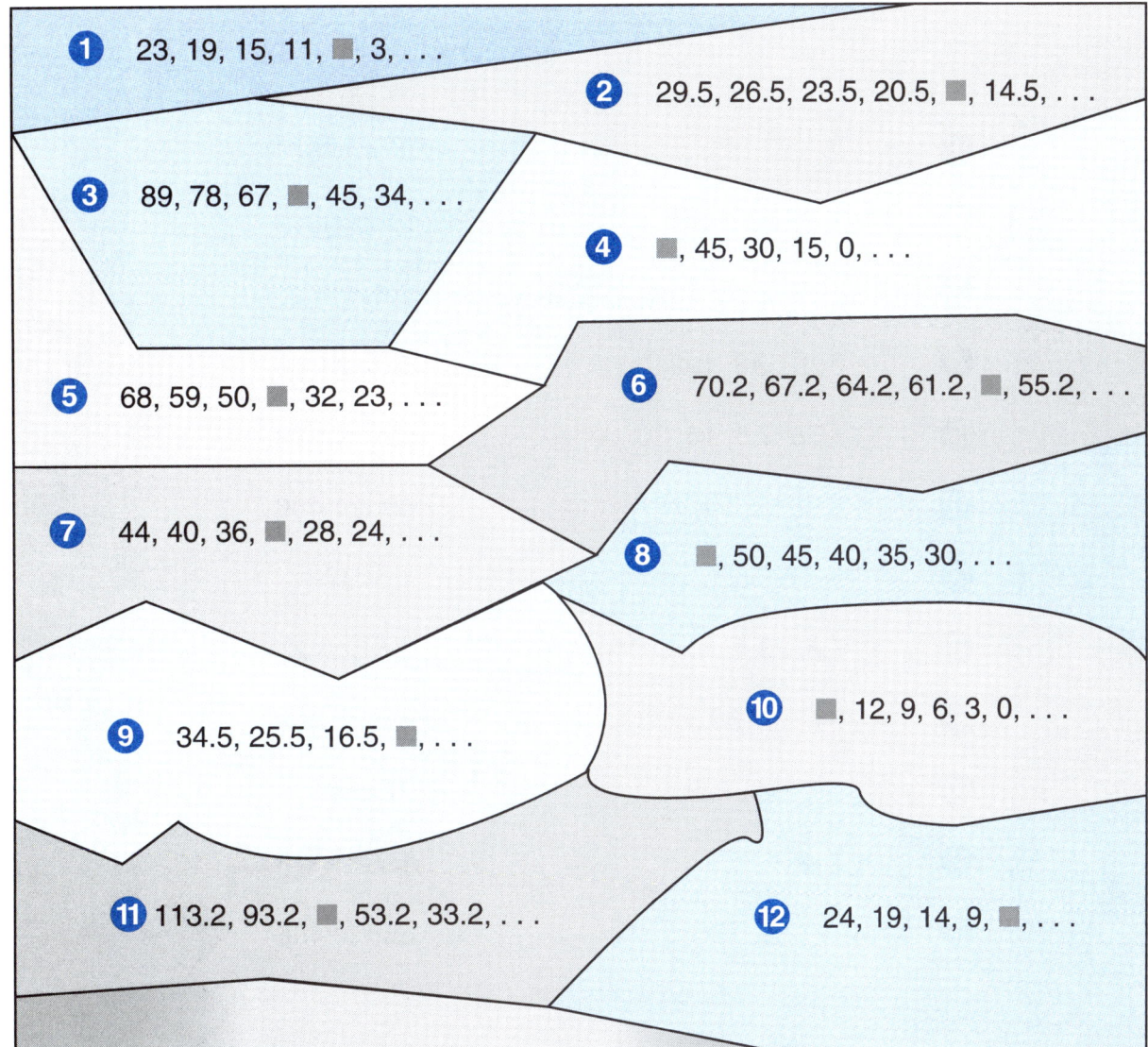

1. 23, 19, 15, 11, ■, 3, . . .
2. 29.5, 26.5, 23.5, 20.5, ■, 14.5, . . .
3. 89, 78, 67, ■, 45, 34, . . .
4. ■, 45, 30, 15, 0, . . .
5. 68, 59, 50, ■, 32, 23, . . .
6. 70.2, 67.2, 64.2, 61.2, ■, 55.2, . . .
7. 44, 40, 36, ■, 28, 24, . . .
8. ■, 50, 45, 40, 35, 30, . . .
9. 34.5, 25.5, 16.5, ■, . . .
10. ■, 12, 9, 6, 3, 0, . . .
11. 113.2, 93.2, ■, 53.2, 33.2, . . .
12. 24, 19, 14, 9, ■, . . .

Answer Box

A	B	C	D	E	F
55	4	60	7	73.2	41
G	H	I	J	K	L
17.5	58.2	32	15	7.5	56

Objective: Identify a missing number in a subtraction number series.

27

Now Multiplication!

Choose the number that is part of the sequence.

1. . . . , 5, 25, 125, . . .   1 or 375
2. . . . , 3, 6, 12, 24, . . .   27 or 96
3. . . . , 12.4, 24.8, 49.6, 99.2, . . .   2.6 or 6.2
4. . . . , 3, 12, 48, 192, . . .   678 or 768
5. . . . , 3.75, 7.5, 15, 30, 60, . . .   120 or 128
6. . . . , 2, 6, 18, 54, 162, . . .   480 or 486
7. . . . , 3, 15, 75, . . .   1 or 375
8. . . . , 7.8, 23.4, 70.2, 210.6, . . .   2.6 or 6.2
9. . . . , 84.75, 169.5, 339, . . .   678 or 768
10. . . . , 81, 243, 729, . . .   27 or 96
11. . . . , 1, 2, 4, 8, . . .   120 or 128
12. . . . , 7.5, 30, 120, . . .   480 or 486

Answer Box

A	B	C	D	E	F
480	2.6	6.2	678	27	375

G	H	I	J	K	L
96	486	128	120	768	1

Objective: Identify a missing number in a multiplication number series.

Can You Tessellate?

Do these shapes tessellate?

1.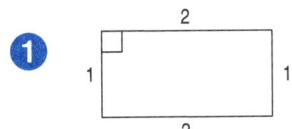
2.
3.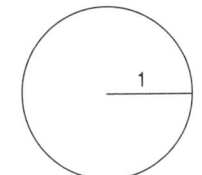
4.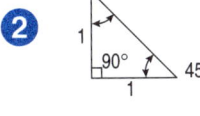
5.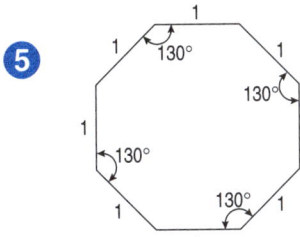
6.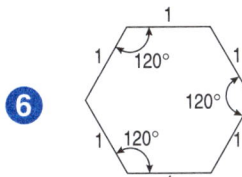

G	yes	A	no
K	yes	F	no
C	yes	J	no
I	yes	H	no
E	yes	C	no
L	yes	B	no

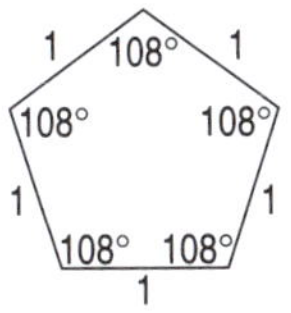

Do these shapes tessellate together?

7.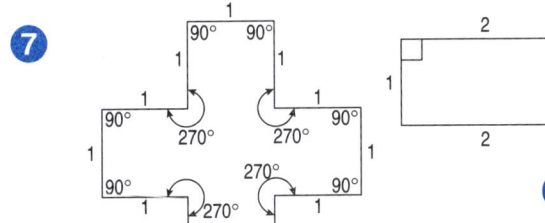
8.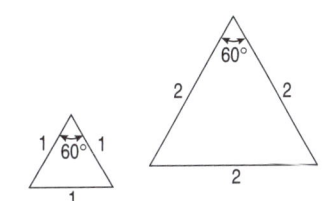
9.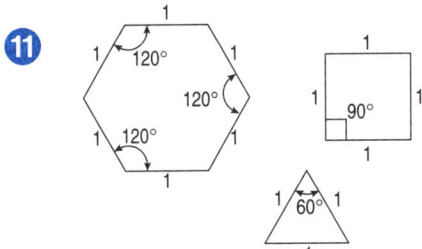
10.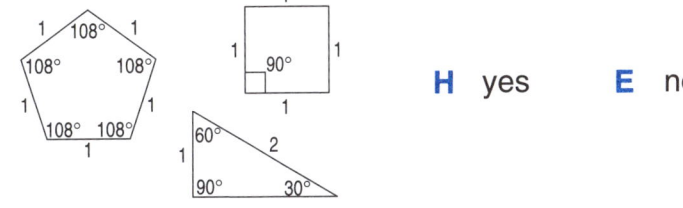
11.
12.

A	yes	D	no
F	yes	K	no
J	yes	I	no
B	yes	G	no
D	yes	L	no
H	yes	E	no

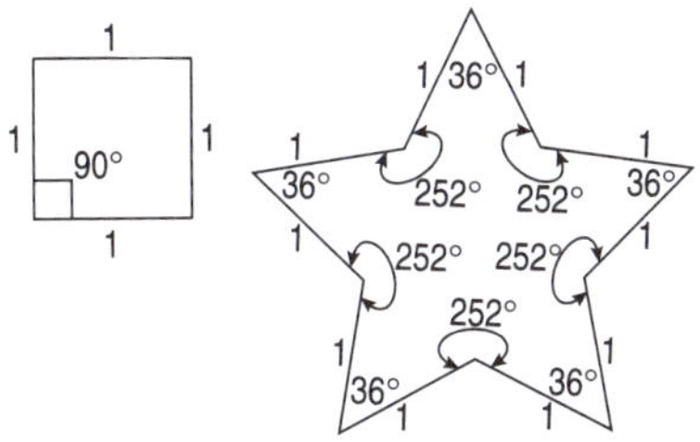

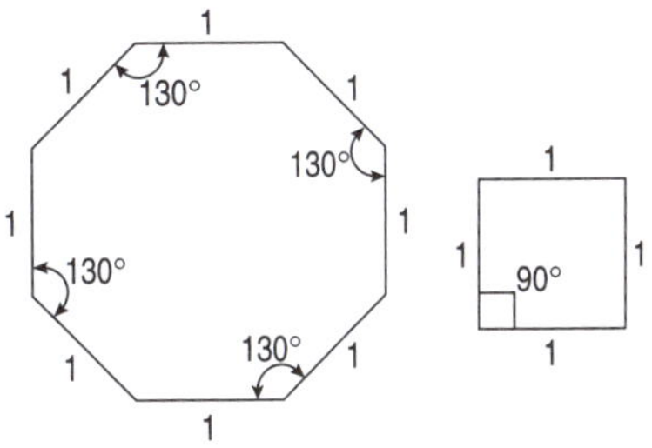

Objective: Identify shapes that tessellate.

29

Pretty Patterns

Find the next figure in the pattern.

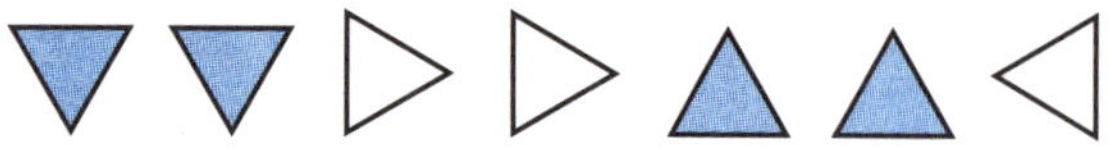

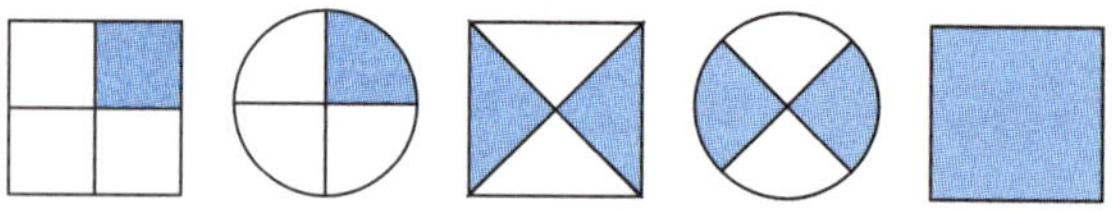

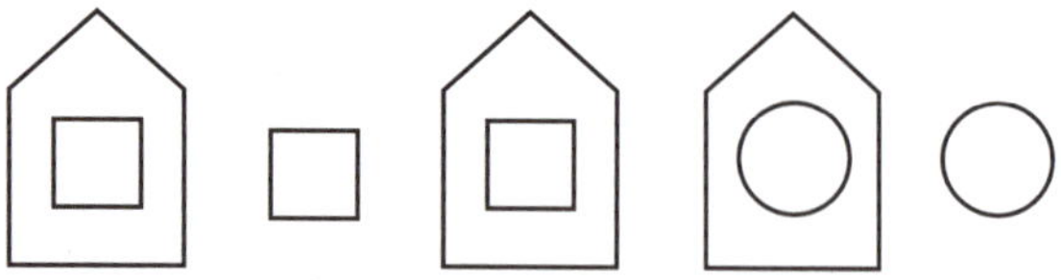

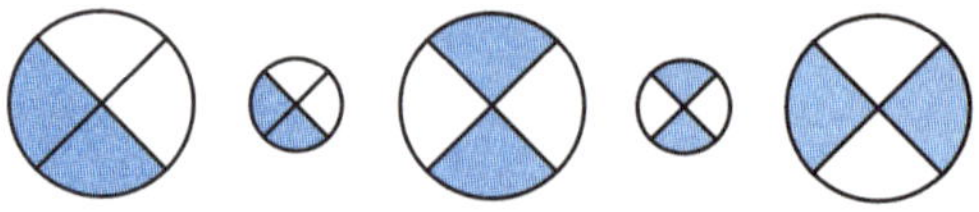

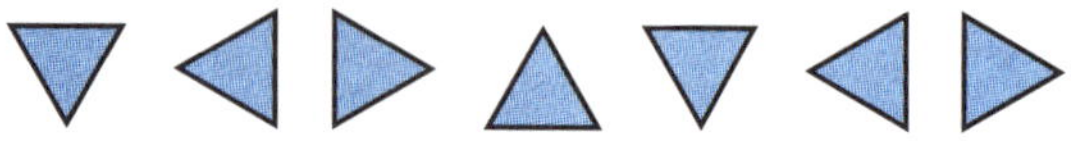

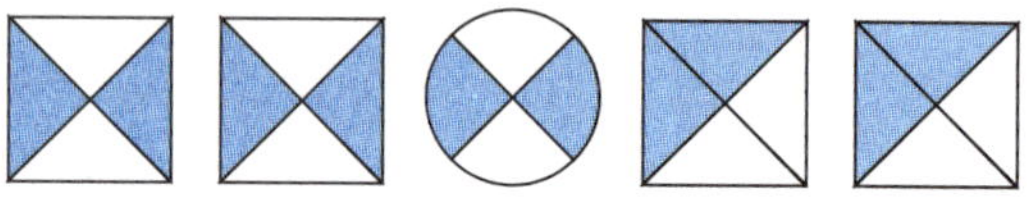

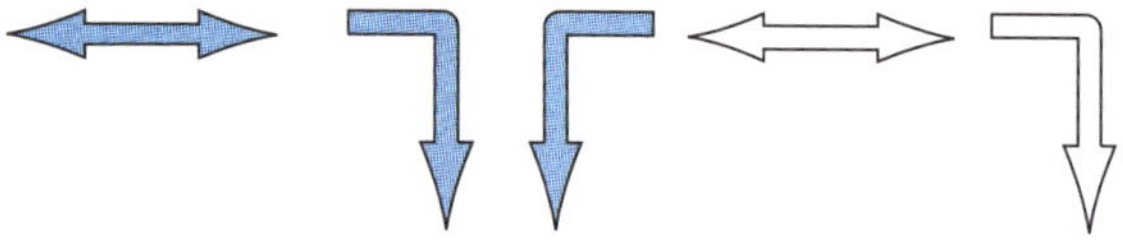

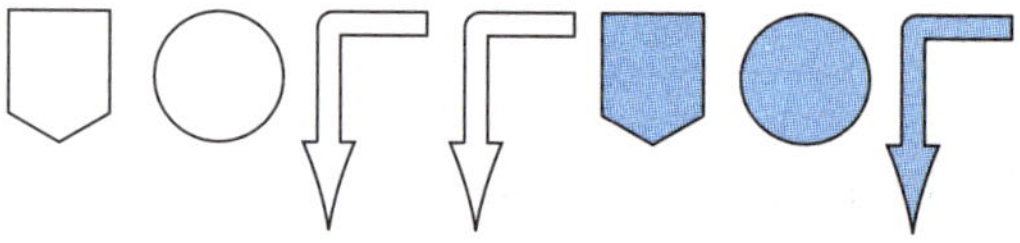

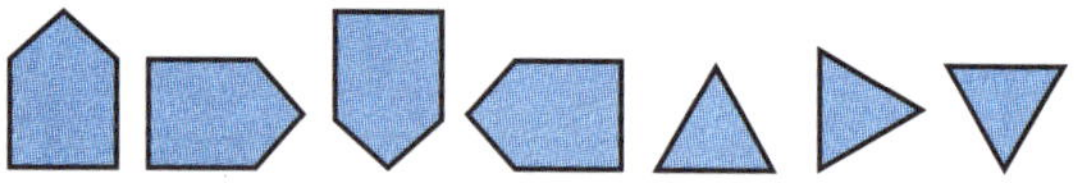

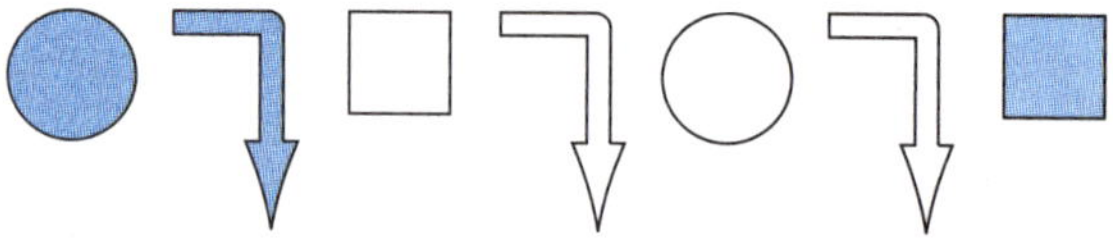

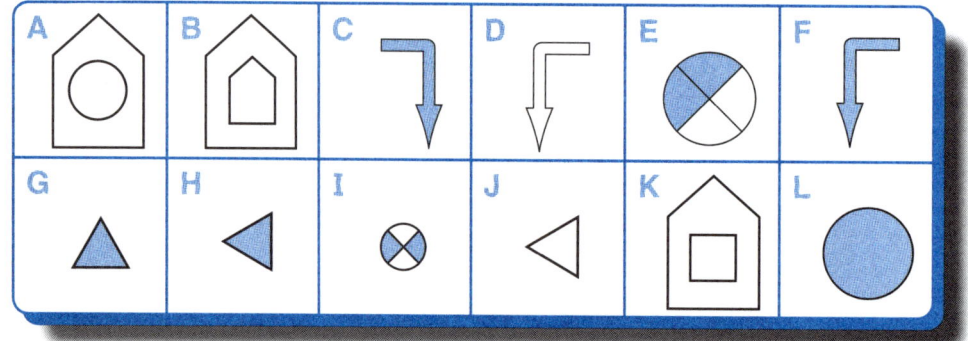

Objective: Determine the next figure in a pattern.

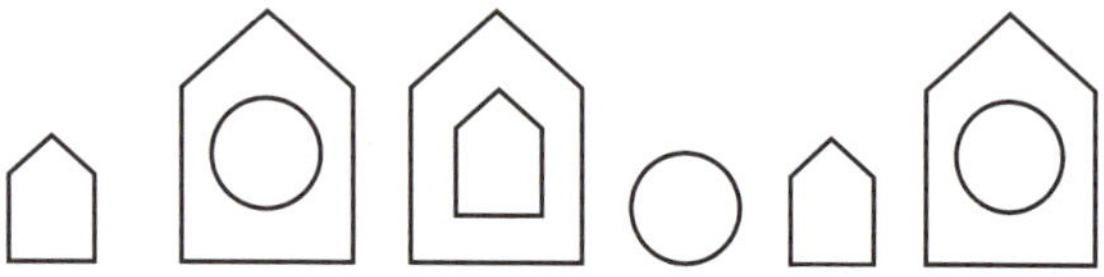

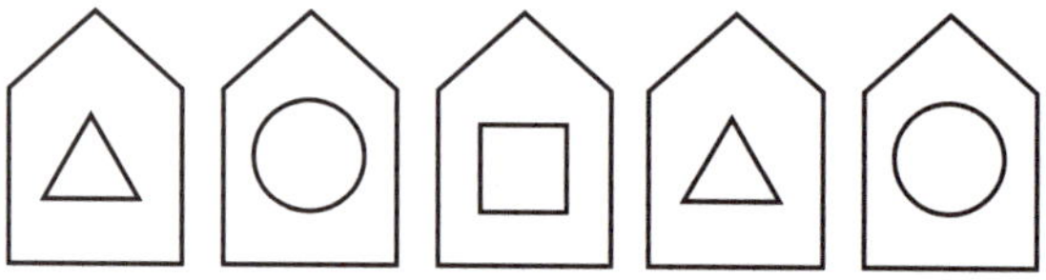

Building Blocks!

Find how many cubes are in the next figure in the pattern.

1

2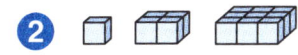

3

4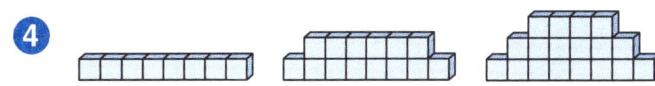

5

6

7

8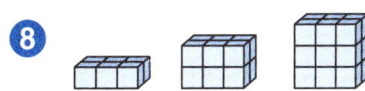

9

10

11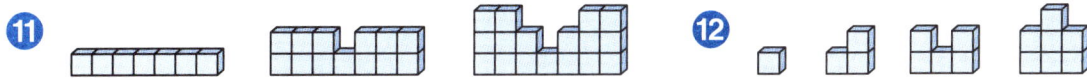

12

Answer Box

A	B	C	D	E	F
30	16	20	64	10	7
G	**H**	**I**	**J**	**K**	**L**
24	9	5	15	19	18

Objective: Extend a building block pattern.

31

Twists and Turns

Find the letter or number that is on the bottom of the cube.

1.
2.
3.
4.
5.
6.
7.
8.
9.
10.
11.
12.

Answer Box

A	B	C	D	E	F
Letter C	6	4	3	Letter A	1
G	**H**	**I**	**J**	**K**	**L**
2	Letter D	Letter F	5	Letter B	Letter E

32 Objective: Identify the face on a given cube after several views have been given.